Mountains

Celia Tidmarsh

BLACKBIRCH®
PRESS

THOMSON

GALE

San Diego • Detroit • New York • San Francisco • Cleveland • New Haven, Conn. • Waterville, Maine • London • Munich

Geography First

Titles in this series
Coasts • Islands • Maps and Symbols
Mountains • Rivers and Streams • Volcanoes

THOMSON

GALE

Produced by White-Thompson Publishing Ltd.
2/3 St. Andrew's Place
Lewes BN7 1UP, U.K.

For more information, contact
The Gale Group, Inc.
27500 Drake Rd.
Farmington Hills, MI 48331-3535
Or you can visit our Internet site at http://www.gale.com

Geography consultant: John Lace, School Adviser
Editor: Katie Orchard
Picture research: Glass Onion Pictures
Designer: Chris Halls at Mind's Eye Design Ltd, Lewes
Artist: Peter Bull

Page 3: The Zuñi sacred mountain of Dowa Yallane in New Mexico.
Page 31: A hill farmer in Greece.

Originally published by Hodder Wayland,
an imprint of Hodder Children's Books,
a division of Hodder Headline Limited
338 Euston Road, London NW1 3BH

Acknowledgements:

The author and publisher would like to thank the following for their permission to reproduce the following photographs:
10 Ecoscene (John Corbett),17 (Andrew Brown), 20 (Andy Binns), 25 (Andy Binns); Hodder Wayland Picture Library *contents page, chapter openers,* 6, 12, 21, 31; Oxford Scientific Films 4 (Colin Monteath), 13 (T.C. Middleton), 14 (Colin Monteath), 19 (Mark Jones), 22 (Marty Cordano), 23 (Lon E. Lauber), 27 (Ben Osborne); Photodisc *cover;* Still Pictures 5 (Calvert-UNEP), 9 (Pascal Pernot), 11 (Roland Seitre), 18 (Cyril Ruoso), 24 (Roberta Parkin), 28 (Galen Rowell); WTPix 3 and 26, 31.

LIBRARY OF CONGRESS CATALOGING-IN-PUBLICATION DATA

Tidmarsh, Celia.
    Mountains / by Celia Tidmarsh.
        p. cm. — (Geography first)
    Summary: An overview of mountains, including how they are formed, how they change over time, what lives and grows there, and what resources and dangers mountains offer mankind.
    Includes bibliographical references (p.) and index.
    ISBN 1-4103-0110-9 (hardback : alk. paper)
    1. Mountains—Juvenile literature. [1. Mountains.] I. Title. II. Series.

GB512.T54 2004
910'.02143—dc22   2003016773

Words in bold **like this** are explained in the glossary on page 30.

Contents

What is a mountain?

A mountain is steep-sided high land that towers more than 984 feet (300 m) above the ground around it. High ground that is less than 984 feet tall is called a hill. The highest point of a mountain is called the **peak** or **summit**.

▼ *This climber is near the top of one of the highest peaks in Antarctica.*

Mountains cover one-fifth of the Earth's surface. Most mountains are found in long chains called **mountain ranges**. Every **continent** has at least one large mountain range. There are even mountain ranges under the sea.

▼ *These mountains in Switzerland form part of a large mountain range called the Alps.*

What makes a mountain?

The Earth's surface is a thin layer of rock called the **crust**. The crust is like a giant jigsaw puzzle, made up of smaller pieces called **plates**. These plates move around very slowly. If they collide with one another, the crust is pushed upwards to form a mountain range.

▼ *The Himalayas were formed when two of the Earth's plates collided.*

A volcano is a special type of mountain. Under the Earth's crust is a thick layer of hot, liquid rock called the **mantle**. Where the plates move apart, some of this liquid rock bursts up through the gap. This forms a **volcano**.

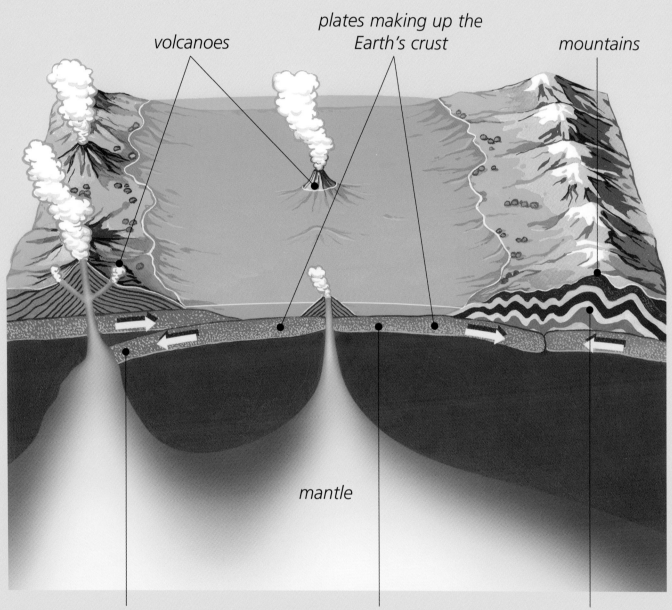

volcanoes

plates making up the Earth's crust

mountains

mantle

An ocean plate is forced under a plate from the land. The ocean plate melts, and hot, liquid rock rises to the surface and forms volcanoes.

Plates move apart under the ocean, and hot, liquid rock rises to form volcanic islands.

Where plates collide, rocks are squeezed together to form mountains.

Fold and block mountains

There are different types of mountains. Where two plates collide, the crust between them is squeezed upwards and folds in on itself. Mountains formed in this way are called **fold mountains**.

Fold Mountains

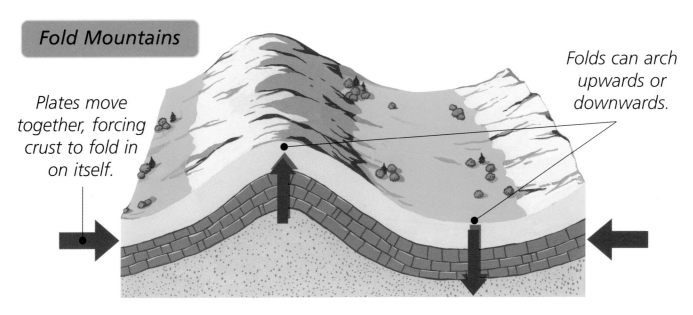

Plates move together, forcing crust to fold in on itself.

Folds can arch upwards or downwards.

Block Mountains

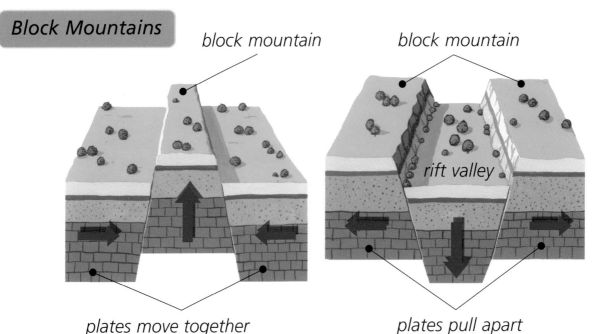

block mountain

block mountain

rift valley

plates move together

plates pull apart

When plates move, large cracks called **faults** appear in the crust. If the crust on either side of two faults moves together, the block of land in between is forced upwards. This forms a flat-topped mountain, called a **block mountain**.

If the crust on either side of two faults moves apart, the land in between slides downwards. This forms a flat-bottomed **valley**, called a rift valley, with block mountains on each side.

▼ *A flat-bottomed valley runs between two block mountain formations.*

Mountains and change

Mountains are constantly changing shape. Once a mountain has been formed, the rock begins to **erode**, or wear away. The hardest rocks take the longest to erode. Erosion can happen in different ways.

All rocks have tiny cracks in them. Ice, wind, rain, and the roots of plants can get into these cracks and make them wider. Eventually pieces of rock will break off and crumble.

▼ *The rocky slope of this mountain in Pakistan has eroded. The crumbly rock is called scree.*

Rivers erode rock, too. Over time, they carve out deeper and deeper valleys through the mountains.

▲ *This river in Bhutan has worn away the rock, forming a steep-sided, V-shaped valley.*

Glaciers

Glaciers are found in many of the world's high mountain ranges. They are like slow-moving rivers of ice.

▼ *This is the Baltoro Glacier in the Karakoram mountains in Pakistan.*

As glaciers move they act like bulldozers. They erode V-shaped river valleys to form deep, broad U-shaped valleys.

Thousands of years ago, in a time called the **Ice Age**, the world was much colder than it is today. There were many more glaciers then. Many of them have melted. But they carved out U-shaped valleys that can still be seen today.

▼ *This U-shaped valley in the mountains in Isère, France, was formed by a glacier.*

Climates

Mountains have different **climates** than low-lying areas. With no surrounding land to shelter it, the peak of a mountain is much colder than the bottom. Mountain peaks are also very windy. High peaks are covered with a blanket of snow all year round. In lower mountain areas, the snow melts when temperatures rise.

▼ *Mountaineers in the Southern Alps of New Zealand struggle against the wind and the cold.*

Some parts of a mountain are always in shadow. They may never get sunshine, and so they are always cool. One side may even have more rain than another. This process is called the **rain shadow** effect.

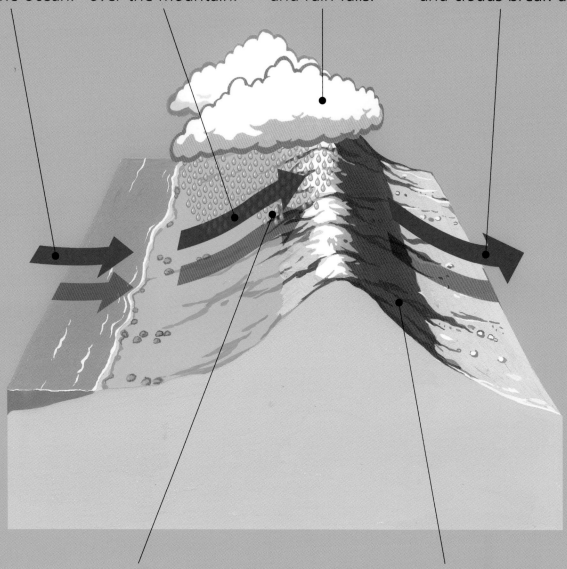

Warm, moist wind blows in from the ocean.

Wind reaches land and is forced over the mountain.

Clouds form, and rain falls.

Wind passes over the mountain and sinks, and clouds break up.

Most of the water carried by the wind falls as rain on the side of the mountain nearer the ocean.

The other side of the mountain has less rain and is called the rain shadow area.

Mountain trees and plants

The types of plants and trees that grow on a mountain depend on its climate.

Deciduous trees need warmth and shelter from the wind. They grow only on the lower mountain slopes. **Evergreen trees** can cope better with the cold and grow farther up the mountain.

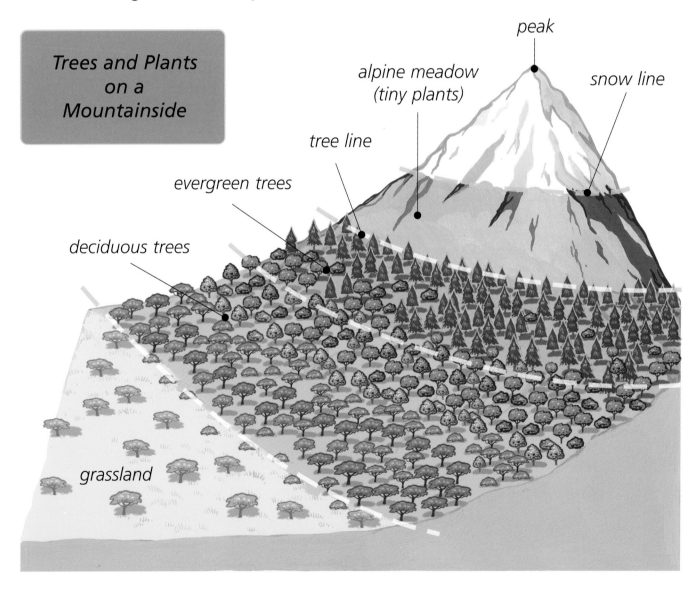

Trees and Plants on a Mountainside

peak

alpine meadow (tiny plants)

snow line

tree line

evergreen trees

deciduous trees

grassland

Higher up, beyond the **tree line**, it is too cold and windy for trees to grow. Only tiny plants can survive here. They have long roots to anchor them to the ground. Nothing grows on the peak because the ground is always covered with snow.

▼ *Tiny flowers grow high above the tree line in Rocky Mountain National Park in Colorado.*

Mountain wildlife

Mountain wildlife has adapted to live in the difficult conditions. Mountain sheep and goats have sharp hooves that can grip the steep, slippery rocks. They also have thick, shaggy coats to keep them warm.

Very few animals stay on the highest slopes all year round. Small animals, such as chipmunks and hares, live there but need to **hibernate** during the winter.

▼ The ibex has a shaggy coat that keeps it warm in cold mountain temperatures.

Larger animals, such as pumas, bears, and wolves live in the forests of the lower slopes, where there is more food.

▲ *Large birds of prey such as condors glide over mountain slopes, hunting for food.*

People of the mountains

People who live in high, remote mountain areas have to adapt to the harsh environment. The steep slopes mean that there is often very little land that can be used for farming. In some places, farmers cut **terraces**, or wide steps, into the mountainside to make room for crops.

▼ Terraces cut into this mountainside in Nepal provide enough flat space for crops to grow.

Steep slopes are difficult to build homes on. It is also difficult to build roads and railways into the mountains. In some mountain communities, people cannot easily get from one place to another and may feel cut off.

▼ *This mountain railway in Argentina provides transport to people who live in remote areas.*

Natural riches

Many of the rocks that make up mountains are useful to people. Granite, limestone, and marble are all rocks that are used for building. **Mines** and **quarries** have been developed in some mountains to reach these rocks.

▼ *This copper mine is in Arizona. Copper is used for making pipes.*

Mining and quarrying involve blasting large areas of rock from the mountainside with explosives. The fragile mountain environment can be badly damaged.

Forestry is another important mountain industry. Forests are often planted on the lower slopes of mountains. They provide wood for building, for making paper, and for fuel.

▼ This logging truck is being loaded with fir trees.

Dangerous mountains

Mountains can be dangerous places. One of the biggest dangers is from **avalanches**. A loud noise may cause huge amounts of snow to slide down the mountainside. Whole villages can be buried under an avalanche. Roads may be damaged, and telephone lines swept away.

▼ An avalanche rumbles down a mountainside in Colorado.

In the mountains the weather can change very quickly. A clear, sunny day may turn chilly suddenly as thick, low clouds move in. The severe cold and strong winds in mountain areas can kill people if they cannot find shelter and keep warm.

▲ *Landslides of crumbling rock can damage roads or even kill people.*

Mountains under threat

Mountain environments may look solid, but they are easily damaged. Many things that humans do can harm the mountain environment.

Large numbers of hikers on a mountain may trample fragile plants and wear away rocks. People may also cause harm by dropping litter.

▼ *People who hike in mountains may harm the environment they have come to see.*

Many countries have opened special protected areas, or **national parks**, in the mountains. People are allowed to visit these areas, but they must not drop litter. They must also stay on posted pathways.

▼ *In this national park in France, researchers monitor and protect mountain wildlife, such as the ibex.*

Mountain fact file

1. The highest mountain on Earth is Mauna Kea. It juts up through the Pacific Ocean as an island in Hawaii. It is 33,481 feet (10,205 m) high.

2. The world's highest mountain on dry land is Mount Everest in the Himalayas (below). It is 29,029 feet (8,848 m) above sea level.

3. The Matterhorn, on the Swiss-Italian border, is a famous example of a sharp-peaked type of mountain, known as a "horn."

4. The highest mountain in Africa is Mount Kilimanjaro, in Tanzania, at 19,340 feet (5,895 m).

5. The Himalayas, in Asia, make up the tallest mountain range on Earth, with nearly 100 of the world's highest peaks.

6. The Andes range is the second-highest mountain range in the world. It runs the length of the continent of South America.

7. The Half Dome in Yosemite, California, has the most vertical mountain face in the world. It drops 2,198 feet (670 m).

8. The Alps stretch across parts of France, Switzerland, Italy, and Austria. They were originally formed millions of years ago under the sea.

9. Australia's highest mountain is Kosciusko in the Great Dividing Range. It is 7,418 feet (2,231 m) above sea level.

10 For many years the Rockies were a barrier between the west coast and the rest of North America. Today railways and roads run across these mountains.

11 Mount Fuji, in Japan, is a volcano that is shaped just like a cone. Mount Fuji has not erupted for hundreds of years.

The World's Major Mountain Ranges

Urals

Rockies

Alps

3 — 8

Himalayas

2

Appalachians

10

7

5

1

Pacific
Ocean

Atlas Mountains

11

Ethiopian
Highlands

East African
Mountains

4

Andes

6

Drakensburg

Great Dividing Range

9

Southern
Alps

Numbers on this map refer to numbers in the fact file.

Glossary

Avalanche A large mass of snow that falls down a mountainside.

Block mountain A flat-topped mountain formed when a block of land is pushed up or when surrounding blocks of land slide down.

Climate The average weather conditions of a place or area over a year.

Continent One of the large pieces of land on the Earth's surface.

Crust The thin, outer layer of the Earth.

Deciduous tree A tree that loses its leaves in autumn.

Erode Wear away.

Evergreen tree A tree that keeps its leaves all year round.

Fault A break or crack in the rocks of the crust.

Fold mountain A mountain formed when the Earth's crust is crumpled up between two plates moving toward each other.

Glacier A river of ice found high up in mountains.

Hibernate To sleep through the winter.

Ice Age A period about 18,000 years ago, when ice sheets covered much of Europe, North America, and Asia.

Mantle The layer of hot liquid rock beneath the Earth's crust.

Mine A place where people have dug out rocks by tunneling underground.

Mountain range A line or chain of mountains.

National park A special area of land that is protected from damage.

Peak The top or highest point of a mountain (also called the summit).

Plate A large section of the Earth's crust that moves about on the mantle.

Quarry A place where people dig out rocks on the surface of the Earth.

Rain shadow The side of a mountain sheltered from the rain.

Summit The top or highest point of a mountain (also called the peak).

Terrace A flat field cut like a step into the side of a mountain.

Tree line The area on a mountain above which trees do not grow.

Valley A large trench that runs between hills and mountains.

Volcano A mountain with a hole or vent through which hot, liquid rock comes up from the mantle, through the crust, and onto the Earth's surface.

For more information

Brimner, Larry Dane. *Mountains.* New York: Childrens Press, 2000.

Durbin, Christopher. *Volcanoes.* San Diego: Blackbirch Press, 2004.

Fowler, Allan. *Living in the Mountains.* New York: Childrens Press, 2000.

Gray, Susan Heinrichs. *Mountains.* Minneapolis, MN: Compass Point Books, 2001.

Pipes, Rose. *Mountains and Volcanoes.* Austin, TX: Raintree Steck-Vaughan, 1998.

Serrano, Marta. *Volcanoes.* San Diego: Blackbirch Press, 2002.

Winne, Joanne. *Living on a Mountain.* New York: Childrens Press, 2000.

Index

All the numbers in **bold** refer to photographs and illustrations as well as text.